# LLAMA AND ALPACA CROSSBREEDING

## A Unique Approach To Camelid Farming

Combine The Best Of Llamas And Alpacas For A Profitable And Unique Livestock Venture

**Dr. Fabian Felicity**

# Table of Contents

# CHAPTER ONE

## Introduction

The world of camelid farming has undergone a remarkable history, highlighted by the hybridization of two separate but closely related species: llamas and alpacas. This convergence, known as crossbreeding, has influenced the history of these South American camelids while also providing several advantages to the agricultural business.

To grasp the importance of llama and alpaca crossbreeding, one must first learn about their history, the distinct qualities and attributes of each species, and the benefits that result from this purposeful hybridization.

## History Of Llama And Alpaca Crossbreeding

The history of llama and alpaca crossbreeding dates back to ancient South American civilizations, when indigenous people tamed these camelids. Llamas (Lama glama) and alpacas (Vicugna pacos) are part of the camelid family, which includes guanacos and vicuñas.

The Inca culture, which flourished in the Andes, understood the unique characteristics of llamas and alpacas and used their powers in many capacities.

Llamas, noted for their strength and stamina, were used as pack animals to transport big goods over the Andes' rough terrains.

Alpacas, on the other hand, were valued for their beautiful, luxuriant fleece, which was used to make wonderful fabrics. The planned mating of llamas and alpacas in these ancient communities established the groundwork for deliberate crossbreeding.

Over the years, selective breeding strategies arose to improve certain features in children. Crossbreeding attempted to produce hybrids that matched the strength of llamas with the desired fiber properties of alpacas. This historical backdrop highlights the cultural, economic, and practical incentives for llama and alpaca crossbreeding, demonstrating how human involvement has affected the genetic landscape of these camelids.

## Understanding Llamas And Alpacas: Characters And Traits

Llamas and alpacas, while having the same ancestor, have diverse qualities and attributes that make each species well-suited for certain tasks in agricultural and textile manufacture.

Llamas are the largest of the two species, having a strong body and distinctive long neck. Their principal use has traditionally been as pack animals, able to transport huge goods over long distances.

Llamas are also recognized for their vigilant and protective attitude, which makes them excellent caretakers of other animals. Their wool, although coarser than that of alpacas, nonetheless has excellent properties and is used in a variety of textile applications.

Alpacas, on the other hand, are smaller and more delicate, valued mostly for their luxuriant fleece. Alpaca fiber is incredibly soft, hypoallergenic, and available in a variety of natural hues. These animals have a mild nature, making them ideal for small farms or as companion animals. While they lack the load-bearing ability of llamas, their fiber quality has established them as important contributors to the textile industry.

The goal of crossbreeding is to blend the strengths of llamas and alpacas. Farmers may develop hybrids that keep the size and power of llamas while inheriting alpacas' fine, high-quality fiber by choosing breeding animals with desirable qualities.

This deliberate mixing of genetic material allows for a more personalized approach to camelid farming, resulting in animals well-suited for a variety of agricultural and commercial applications.

## Advantages Of Crossbreeding In Camel Farming

The purposeful crossbreeding of llamas and alpacas provides several advantages that improve the overall efficiency and production of camelid farming operations.

1. Fiber Quality and Quantity: Crossbreeding enables producers to create animals with a mix of alpaca fiber softness and llama wool quantity. This dual-purpose method meets the needs of both the textile

industry and agricultural uses, resulting in a more adaptable and economically sustainable herd.

2. Size and Strength: Llamas, with their bigger size and strength, add to the genetic mix by passing on these characteristics to hybrid offspring. This is especially beneficial for farmers who need pack animals or who live in areas with difficult terrain where the strength of llamas is essential.

3. flexibility: Crossbred llama-alpaca hybrids often demonstrate flexibility, making them suitable for a wide range of temperatures and situations. This versatility is due to the different ecological niches that llamas and

alpacas occupy in their natural South American environments.

4. illness Resistance: Intentionally combining genetic material may improve the general health and illness resistance of the hybrid progeny. This is critical for maintaining a robust and flourishing herd, as it reduces susceptibility to common diseases that may afflict both llamas and alpacas.

5. Market Demand: The crossbreeding technique addresses market demand for animals that can provide numerous functions. Whether it's producing high-quality fiber or using llamas as pack animals and guards, crossbred camelids can

meet a wide range of agricultural and economic purposes.

To summarize, the history and practice of llama and alpaca crossbreeding demonstrate the complex interaction between humans and these South American camelids. From their importance in ancient civilizations to the deliberate incorporation of their genetic material into current agricultural operations, llamas and alpacas continue to play important roles in agriculture and industry. The deliberate coupling of these species, motivated by a desire to maximize favorable features, has produced hybrids with a balanced mix of size, strength, and fiber quality. As the camelid agricultural environment changes, the heritage of crossbreeding lives on, forming herds that are not only commercially useful but also adaptive to the changing needs of modern agriculture.

# CHAPTER TWO
## Choosing Ideal Breeding Pairs
## For Crossbreeding

Crossbreeding llamas and alpacas is a precise procedure that must take into account a variety of elements to produce healthy and acceptable offspring.

Selecting optimal breeding partners is an important stage in this procedure since it directly affects the genetic composition and traits of the crossbred animals.

When selecting breeding couples, it is important to consider the specific characteristics of both llamas and alpacas. This includes evaluating their conformation, fiber quality, temperament, and general health. Conformation refers to an animal's structural correctness, which includes body proportions, leg alignment, and general physical health. Breeders may increase their chances of generating crossbred offspring with better structural integrity by choosing individuals with ideal conformation.

Fiber quality is another important factor to consider, particularly for those who value producing high-quality fibers in their crossbreeding program. Llamas and alpacas are prized for their rich fibers, and breeders strive to mix the finest features of both species to create the ultimate combination. This entails analyzing the fibers' fineness, length, and color, to strike a balance that matches market needs while also meeting the breeder's objectives.

Temperament is an important consideration in breeding choices since it affects the ease of handling and general management of crossbred animals. Llamas are noted for their protective and alert behavior, but alpacas are often more placid.

Combining these features intelligently may result in crossbred offspring with temperaments that are appropriate for the breeder's desired goal, whether it is as a guardian animal, a fiber producer, or a companion.

In addition to individual qualities, genetic compatibility is an important consideration when choosing breeding pairings. Breeders must understand the genetic heritage of both llamas and alpacas, including recessive features, possible genetic illnesses, and general genetic variety within each species.

Breeders may reduce the chance of unwanted characteristics showing in crossbred progeny by thoroughly studying pedigrees and undertaking genetic testing as needed.

Successful crossbreeding must also take into account the breeding program's aims and objectives. Whether the aim is to improve fiber quality, enhance conformation, or generate animals for particular reasons, such as therapy or trekking, breeding choices must be aligned with these objectives.

This deliberate strategy assures that each generation of crossbred animals makes a good contribution to the overall progress of the breeding program.

## The Art Of Llama And Alpaca Genetics

Understanding the complexities of llama and alpaca genetics is critical for successful crossbreeding. Both species have distinct genetic features, and a thorough understanding of these qualities is essential for making sound breeding choices.

Llamas and alpacas have a similar ancestor, however, they vary in terms of chromosomal counts and other critical genetic markers. Llamas have 35 pairs of chromosomes, while alpacas have 36. This difference complicates the crossbreeding process, forcing breeders to carefully traverse the genetic landscape to produce effective results.

One important aspect is the possibility of hybrid vigor, commonly known as heterosis, in crossbred llamas and alpacas. Hybrid vigor often produces offspring with improved characteristics such as greater growth rate, disease resistance, and general vitality. However, the level of hybrid vigor varies, and not all crossbred individuals may benefit from these characteristics.

Another concern with crossbreeding operations is inbreeding depression. Breeding closely related people increases the risk of expressing detrimental recessive genes, resulting in worse health and performance in the children. To reduce this danger, breeders must carefully organize matings, avoiding near relatives, and ensuring genetic diversity in the breeding pool.

The use of genetic technology, such as DNA testing, has become an essential element in current breeding operations.

DNA testing enables breeders to detect particular genetic markers, analyze the relatedness of possible breeding partners, and make educated choices to attain the desired results. This method is especially effective for detecting carriers of genetic diseases and ensuring appropriate breeding practices.

# CHAPTER THREE
## Breeding Management And Reproductive Considerations

A successful crossbreeding program involving llamas and alpacas relies on effective breeding management. A well-managed program requires an understanding of both species' reproductive physiology as well as the implementation of effective breeding procedures.

Llamas and alpacas are induced ovulators, which means they do not have a typical estrous cycle. Females ovulate in reaction to mating or other stimuli. Breeders must closely watch females for symptoms of heat and use strategic breeding strategies to maximize conception rates due to this distinct reproductive characteristic.

Artificial insemination (AI) is a widespread procedure in llama and alpaca breeding, and it has various benefits, including the ability to utilize genetic material from animals in different regions and to precisely schedule inseminations for the best outcomes.

However, effective AI requires specialized expertise and technology, and breeders must understand both species' reproductive systems.

Managing the pregnancy and birthing process is also essential in a crossbreeding program. Proper diet, veterinary treatment, and monitoring are all necessary to guarantee the health and well-being of pregnant women and their kids.

Llamas and alpacas have a gestation period of around 11 months, and recognizing the signals of imminent parturition is critical for prompt intervention if necessary.

Breeding management also considers the postnatal care and growth of crossbred children. Providing adequate feeding, vaccines, and socialization improves the animals' general health and temperament, laying the groundwork for their future responsibilities in the breeding program.

# Health And Nutrition Of Crossbred Llamas And Alpacas

The health and nutrition of crossbred llamas and alpacas are critical to their survival and long-term production. A proactive healthcare strategy and a well-balanced dietary program are critical components of responsible breeding and maintenance.

Regular veterinarian treatment is essential for preserving the health of crossbred llamas and alpacas. Preventive procedures including vaccines, parasite management, and dental care serve to reduce illness risk and maintain the animals' general well-being. Close monitoring for indicators of disease or discomfort enables quick management, minimizing the possible effect of health concerns on the breeding effort.

Nutritional needs vary depending on the age, reproductive condition, and planned usage of the crossbred animals. A balanced diet that addresses these individual demands is essential for good health and performance. Llamas and alpacas are noted for their effective foraging abilities, although vitamin and mineral supplementation may be required, particularly in areas with nutrient-deficient food.

Crossbred llamas and alpacas' health is also heavily influenced by water quality and availability. Clean and readily available water sources are critical for optimal digestion, thermoregulation, and general hydration. Breeders must ensure that their animals have access to enough water at all times, particularly during hot weather or when breastfeeding.

To summarize, identifying appropriate breeding couples for crossbreeding llamas and alpacas requires a thorough examination of individual features, genetic compatibility, and alignment with breeding objectives. Understanding the genetic nuances of both species is critical for making educated judgments and negotiating the complexity of hybrid vigor and inbreeding depression.

Effective breeding management, including the use of reproductive technology, is critical to attaining positive results. Finally, focusing on the health and nutrition of crossbred llamas and alpacas promotes their well-being while also contributing to the breeding program's long-term success.

# Challenges In Crossbreeding: Identifying And Overcoming

Crossbreeding, or marrying individuals from different breeds or species, has grown in popularity in the animal husbandry industry for a variety of reasons. In the case of llamas and alpacas, crossbreeding offers both benefits and problems.

Identifying and overcoming these problems is critical for effective crossbreeding initiatives, especially in terms of economic sustainability, marketing tactics, and new methodologies.

Crossbreeding llamas and alpacas presents unique problems, one of which is assuring the health and well-being of the hybrid progeny. While these two South American camelids have the same ancestor, their variances in size, wool qualities, and temperament make mating difficult. Mating bigger llamas with smaller alpacas, for example, might lead to problems during pregnancy and delivery.

To achieve successful pregnancies and healthy kids, these hurdles must be overcome by precise breeding partner selection and, in certain situations, assisted reproductive technology.

Another issue with crossbreeding is the possible dilution of good features.

Llamas are often appreciated for their size, strength, and aptitude as pack animals, but alpacas are esteemed for their rich, luscious fiber. In the goal of generating a hybrid with a combination of these properties, there is a danger of losing the unique characteristics that make each species valuable.

Breeders must carefully balance selection criteria to maintain good features while also adding new and complementing attributes to hybrid offspring.

Addressing the issue of genetic variety is equally important in crossbreeding operations. Overemphasis on certain features may result in a narrow gene pool, raising the incidence of genetic illnesses and decreasing the population's general resilience. Breeders must employ techniques to preserve genetic variety within crossbred populations, which will ensure the hybrid animals' long-term health and adaptability.

# CHAPTER FOUR
## Economic Viability Of Llama-Alpaca Crossbreeding

The economic feasibility of crossbreeding llama and alpaca populations is an important factor for breeders and industry players.

Assessing the cost-effectiveness and possible returns on investment requires a thorough knowledge of both the market demand for hybrid camelid goods and the efficiency of the crossbreeding process.

One facet of economic viability is the capacity of crossbred llamas and alpacas to serve several functions. For example, a hybrid camelid may have the bigger size and power of llamas combined with the fine fiber properties of alpacas. This dual-purpose nature creates chances for a variety of revenue streams, including meat production, pack animal services, and high-quality fiber manufacturing.

However, successful exploitation of this potential requires competent breeding procedures, efficient management, and a responsive market.

Crossbreeding initiatives' economic performance is heavily influenced by market demand. Understanding customer preferences for certain characteristics, such as fiber quality or size, is critical for adjusting breeding efforts to market demands. Furthermore, teaching customers about the distinct characteristics of crossbred camelid goods has the potential to establish specialized markets and increase breeders' profits.

In this environment, challenges include the ever-changing nature of customer tastes and the need for ongoing adaptation to market trends.

Managing production costs is also necessary to ensure crossbreeding's economic viability. Feed, veterinary care, and infrastructural fees all add considerably to the total cost of camelid rearing. Breeders must find a balance between investing in the animals' welfare and running a cost-effective company. Innovations in feed formulas, healthcare procedures, and facility architecture may help to reduce production costs and improve the overall economic viability of crossbreeding projects.

## Marketing Strategy For Crossbred Camelid Products

Effective marketing of crossbred camelid goods is crucial to guaranteeing the success of crossbreeding initiatives. The unusual mix of features in hybrid llamas and alpacas provides a chance for market distinctiveness, but it also requires tailored efforts to convey these benefits to customers.

Highlighting the dual-purpose character of crossbred camelids is a key marketing strategy. Emphasizing the adaptability of these hybrids for meat and fiber production may attract a larger customer base.

Marketing efforts should inform prospective purchasers of the advantages of owning crossbred llamas and alpacas, whether as pack animals, friends, or suppliers of high-quality fiber. Creating a good and captivating story about hybrid animals may help to establish a strong market presence.

Building industry ties and cooperation may help to boost marketing efforts for crossbred camelid goods. Engaging with fiber artists, fashion designers, and other textile industry stakeholders may open up new opportunities to showcase the distinct features of crossbred llama and alpaca fiber.

Collaborative marketing initiatives that promote the ecological and eco-friendly qualities of crossbred goods may appeal to environmentally concerned customers, broadening the market reach for these hybrids.

Online platforms and social media play a key part in current marketing techniques. Breeders should exploit these media to promote their crossbred camelids, share success stories, and communicate with prospective buyers.

The visual attractiveness of llama and alpaca hybrids, along with an interesting narrative, may build a strong online presence, drawing a varied audience interested in rare and different creatures.

# CHAPTER FIVE
## Innovations In Llama And Alpaca Crossbreeding Techniques

Innovations in crossbreeding methods are at the forefront of overcoming problems and boosting the effectiveness of hybridization initiatives.

Advancements in assisted reproductive technologies, genetic engineering, and breeding methods help to improve the process of developing crossbred llamas and alpacas with desired qualities.

Assisted reproductive methods, such as artificial insemination and embryo transfer, are critical in overcoming the size and temperament disparities between llamas and alpacas. These technologies allow breeders to carefully choose mating partners, regulate breeding cycles, and maximize the chance of successful births. The accuracy provided by assisted reproductive methods improves the overall effectiveness of crossbreeding projects.

Genetic engineering shows potential for modifying certain features in crossbred populations. While ethical issues and public approval are vital, advances in genetic technologies may help select and improve desirable traits. Genetic markers for fiber quality, disease resistance, and other commercially useful features may be found and used in crossbreeding programs to speed up the generation of better hybrid animals.

Selective breeding methods remain an important part of crossbreeding innovation. Breeders are constantly refining their selection criteria based on performance data, market trends, and customer feedback.

The combination of technology like as genomics and phenomics provides for a more thorough knowledge of individual animals' genetic potential, allowing breeders to make educated choices in the pursuit of desired qualities.

To summarize, the problems of crossbreeding llama and alpaca populations are varied, including health concerns, economic sustainability, marketing methods, and creative breeding procedures. Overcoming these issues requires a comprehensive strategy that combines the preservation of good characteristics with the introduction of new ones.

By resolving these issues, breeders may maximize the economic potential of crossbred camelids, resulting in viable and marketable hybrid populations that contribute to the llama and alpaca industries' variety and resilience.

# Environmental Considerations For Crossbred Camelid Farming

Crossbred camelid farming is the deliberate breeding of multiple species or kinds of camelids to accomplish particular objectives such as increased production, disease resistance, or adaptability to diverse habitats.

While the emphasis is often on the economic advantages of crossbreeding, it is critical to address the environmental consequences of such activities. This section investigates the environmental aspects of crossbred camelid farming, offering information on how these practices affect ecosystems and biodiversity.

One of the main environmental concerns is the possible damage to local ecosystems. When crossbred camelids are brought into new areas, they may outcompete local species for resources, resulting in a decrease in biodiversity.

This emphasizes the significance of performing comprehensive environmental impact evaluations before starting crossbreeding initiatives. Understanding the possible ecological repercussions may assist farmers and governments in making sound choices that strike a balance between economic rewards and environmental sustainability.

Furthermore, the excrement produced by crossbred camelids might harm the surrounding ecology. To avoid pollution of land and water, proper waste management methods must be used.

This involves devising ways for properly disposing of feces and other byproducts of crossbred camelid farming. Implementing sustainable waste management procedures preserves the environment while also ensuring the long-term success of crossbreeding operations.

# Case Studies: Successful Crossbreeding Ventures

Examining case studies of successful crossbreeding initiatives will help you understand the practical issues of putting these programs into action.

Successful crossbreeding projects include not just economic and genetic aspects, but also their environmental effects. One famous example is the crossbreeding of alpacas with dromedary camels in a semi-arid area.

In this example, the objective was to develop a camelid hybrid that could flourish in dry environments while retaining desirable characteristics like wool quality and reproductive efficiency. The success of this enterprise was dependent on the careful selection of parent stock, which took into account both genetic compatibility and adaptation to the local climate. The resultant crossbred camelids were more resilient to drought, making them ideal for the region's unique ecological difficulties.

Another example is the crossbreeding of llamas with guanacos in steep terrain. The goal was to create a hybrid camelid with the strength and endurance of a llama and the agility and adaptability of a guanaco. This successful crossbreeding enterprise not only helped the lives of local farmers but also highlighted the possibility of sustainable agriculture in difficult terrains.

These case studies demonstrate the need to customize crossbreeding programs to each region's specific environmental variables. Successful ventures incorporate ecological considerations, ensuring that the resulting crossbred camelids benefit the local ecosystem rather than causing harm.

## Future Prospects And Trends In Camel Farming

As crossbred camelid farming evolves, various future possibilities and trends are influencing the sector. One noteworthy development is the incorporation of genomic technology into breeding efforts. Advances in genetic research enable farmers to make more accurate and educated breeding selections, expediting the development of desired features in crossbred camelids.

Furthermore, the desire for environmentally friendly and ethically produced fibers is encouraging innovation in camelid farming. Consumers are becoming more aware of the environmental and ethical consequences of their purchases, resulting in a rising demand for items derived from properly produced and maintained crossbred camelids. This trend provides farmers with a chance to not only fulfill market needs but also contribute to the agricultural industry's overall sustainability objectives.

Climate-resilient crossbred camelids are anticipated to gain attention in the future years, as the consequences of climate change become more obvious. Farmers may consider crossbreeding projects that try to improve camelids' adaptive skills, assuring their continuous production in the face of changing environmental circumstances. This proactive strategy is consistent with the overall objective of establishing agricultural systems that can withstand the difficulties provided by a fast-changing environment.

## Conclusion

Finally, crossbred camelid farming presents both economic opportunities and environmental concerns. To guarantee the long-term viability of these initiatives, environmental considerations must be carefully considered. Case studies demonstrate that effective crossbreeding projects include ecological factors in their breeding operations, which benefits local ecosystems.

The future of camelid farming will most certainly be dictated by advances in genetic technology, a rising demand for sustainable goods, and the need for climate-resilient varieties. By embracing these developments and prioritizing environmental sustainability, the camelid farming sector may grow while reducing any ecological consequences.

As we manage the intricate interaction between crossbreeding, environmental concerns, and future trends, farmers, academics, and legislators must cooperate in developing solutions that balance the advantages of crossbreeding with the preservation of our natural ecosystems.